Gateways to Science

Gateways to Science

Neal J. Holmes
Department of Chemistry
Coordinator of Science Education
Central Missouri State University

John B. Leake
Department of Science Education
University of Missouri

Mary W. Shaw
Department of Education
Montgomery County, Maryland

Webster Division, McGraw-Hill Book Company
New York/St. Louis/Dallas/San Francisco/Auckland/Bogotá/Guatemala
Hamburg/Johannesburg/Lisbon/London/Madrid/Mexico/Montreal/New Delhi
Panama/Paris/San Juan/São Paulo/Singapore/Sydney/Tokyo/Toronto

Consultants and Reviewers

Patricia Eastland Greenfield Union Elementary, Detroit, Michigan
Alden Loomis Montebello Unified School District, Montebello, California
Richard J. Merrill Mt. Diablo Unified School District, Concord, California
Dr. John P. Huntsberger Science Education Center, The University of Texas at Austin, Austin, Texas
Dr. Harold C. Jorgensen Professor of Education, Portland State University, Portland, Oregon
Miram R. Buckland K. T. Murphy School, Stamford, Connecticut
Joseph T. Piazza Easton Area School District, Easton, Pennsylvania
Rita Hawkins Riverbend School, St. Louis, Missouri
Alice Kimbler Hankla The Galloway School, Atlanta, Georgia
Charles J. LaRue Montgomery County Schools, Rockville, Maryland
Barbara Wolczak St. Joseph Franciscan, Cleveland, Ohio

Editor in Chief: John F. Mongillo
Sponsoring Editor: M. Jane Kita
Editing and Styling: Caroline Levine
Photo Editor: Suzanne V. Skloot
Production: Angela Kardovich
Editorial Assistant: Joann McPartland

Editor: Linda Nicholson
Photo Researcher: Randy Matusow
Text and Cover Design: Group Four, Inc.
Design Supervision: Frank Medina
Cover Photo: H. R. Uthoff/The Image Bank
Layout and Art Production: Cover to Cover, Inc.
Artists: Michael Adams, Roberta Collier, Kathy Hendrickson, John A. Lind Corp.
Editorial Assistant: Lorraine Rogers

This book was set in 16 point Aster by Black Dot. The color separation was done by Black Dot.

ISBN 0-07-029822-X

4 5 6 7 8 9 10 DODO 91 90 89 88 87 86 85 84

Contents

Life science

11 Water safety 79

Earth and space science

12 Rocks 89

13 Life long ago 95

14 Clouds 103

15 Weather 111

16 Up in the sky 119

Physical science

1

Life science

Health and safety

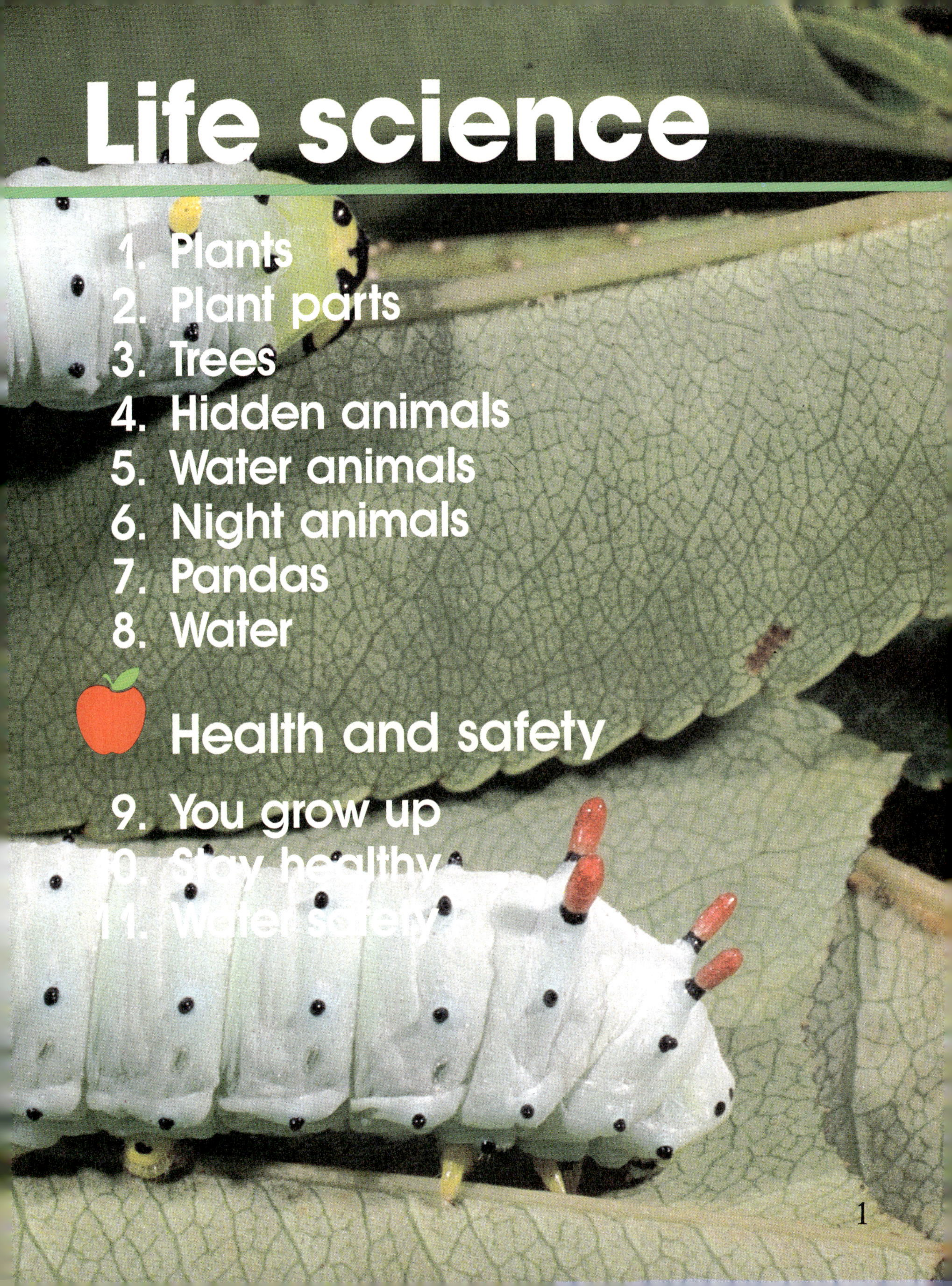

1

Plants

Many kinds of plants

Plants grow in all shapes, colors, and sizes. Some are green. Some are tall. And some are so small they can hardly be seen at all.

Plants are living things. A flower is a plant. Grass is a plant. Name some plants you know. What plants do you see here?

Plants are like people in some ways. How? Both plants and people need special things to grow.

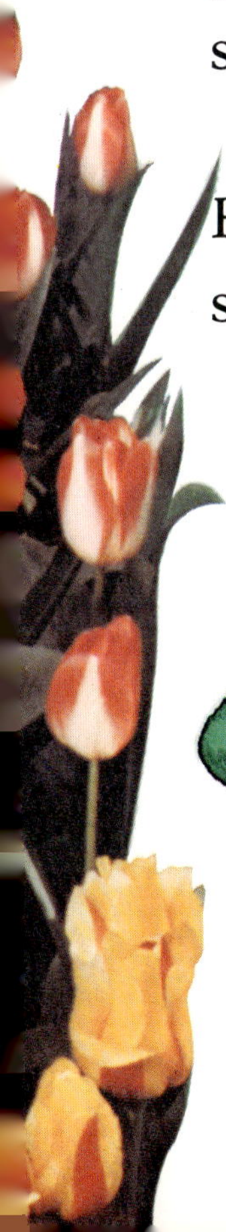

Plants need food

All living things need food to grow. How do you think plants get the food they need?

Plants grow only in places where they can get food. Some plants grow in water. Some plants grow in **soil.** And some plants even grow on other plants. How does the plant in this picture get food? 1

1

Activity

Plants need water

All plants need water to live and grow. Most plants get water from rain. But there are other ways that plants get water. 2 Can you think of some ways?

2

How can you tell that plants need water? Put soil in three empty milk **cartons**. 3 Put several bean seeds in the soil in each carton. Mark two of the cartons with the letter W. Water these cartons each day. Do not water the third carton. What happened to the seeds you watered? What happened to the seeds you did not water?

3

Plants need sunlight

The sun helps most plants grow. How? It gives plants light and warmth.

You can see how the sun helps plants grow. Take your two bean plants that are green and growing. Put one plant in a dark place. Put the other plant near a window. Water **both** plants every day. What happens? Do the plants look different? 1, 2 Why?

1

2

Show what you know

Number your paper from 1 to 6. Write T for each sentence that is true. Write NT for each sentence that is not true.

1. Plants come in all shapes and sizes.
2. Plants are living things.
3. Plants are not like people at all.
4. Plants get food from the place where they grow.
5. Plants need water.
6. Plants grow well in the dark.

Think it through

How are plants like you?

2

Plant parts

Many parts of plants

Are all plants **alike**? They may not look alike, but plants are alike in many ways. Plants have many parts that are the same. Can you name some plant parts?

Parts of plants are used in many ways. Some parts are pretty to see and smell. Some parts are good to eat. And some are used to make things.

What parts of the plants pictured here are the same? 1

1

Plants have roots

Most **roots** are **underground.** Some roots grow long and deep. And some grow fat and round. How are the roots in these pictures alike? 1, 2 How are they different?

What do roots do? Roots help to hold plants in place. And they take food and water from the soil to feed the plants. Let's watch roots grow. Then you will know what they do.

1

2

Activity

Watching roots grow

Soak a bean seed in water to make it soft. 3 Then, place the seed in a glass. Use a wet **blotter** to hold the seed against the side of the glass. Put a little water in the bottom of the glass. The blotter will soak up the water. This will keep the seed wet.

Watch the roots begin to grow. Do the roots grow down? How long are the roots?

3

Plants have stems

Do all plants have **stems**? Are all stems the same? Most plants have stems that grow straight above the ground. 1 But some plants have stems that grow along the ground.

How do stems help plants? Let's find out. Fill a glass with water. Add a few drops of food coloring. 2 Cut off a **stalk** of celery, and place it in the colored water. 3 Look at the stem with your hand lens. What happens? 4

1

2

3

4

Plants have leaves

Leaves are important parts of plants. Leaves come in all shapes and sizes. Leaves may be pretty. And leaves may smell good. But why are leaves important?

Leaves use sunlight, air, and water to make food for plants.

Can you name the plant leaves in this picture? **5** How are the leaves alike? How are they different?

5

Plants have flowers and seeds

What is the prettiest part of a plant? The flower is the prettiest part of a plant.

Flowers are important for plants, too. Most of a plant's seeds grow in the flower.

Can you name the flowers pictured here? 1 Is your favorite flower here? How are these flowers often used?

1

Show what you know

Number your paper from 1 to 5. Read each sentence below. Write T on your paper for each sentence that is true. Write NT for each sentence that is not true.

1. Stems help hold plants in place.

2. Roots help take food and water from the soil.

3. Stems usually grow under the soil.

4. Leaves help plants make food.

5. Most of a plant's seeds grow in the flower.

Think it through

How many different ways do you use plants every day?

3

Trees

How big are trees?

Trees are the world's biggest plants. How big are trees? **California redwood** trees are some of the world's tallest trees. Some have grown more than 110 meters (about 360 feet) tall.

Some kinds of trees grow very fast. Some kinds grow slowly. But all trees keep growing.

A tree's **height** depends on many things. How old is the tree? What kind of soil is it growing in? How much sun and water does the tree get? What is the weather like where it grows?

1

What can you tell about the weather where these trees grow? 1, 2 Which trees lose their leaves in cold weather? Which trees grow in a **jungle**? Which trees grow where it is cold?

2

How old are trees?

Some of the world's oldest plants are trees. Did you know that some **bristlecone pine** trees are 4600 years old. How old is the tree in this picture? 1 How can you tell?

There are circles inside the **tree trunk**. 2 The circles are **tree rings**. A tree grows one ring each year. Count the rings. How many years did the tree grow?

This is the way scientists tell a tree's age. They count the tree rings.

1

2

3 4

Helping the forest

A **forest** is a place where many trees grow.

What happened to the forest in this picture? 3

What happened here? 4

What happened to these trees? 5

Trees are important to us. Wood comes from trees. People need wood. Why is it important for us to **protect** forests?

5

Activity

Grow a tree

Put some soil in an empty milk **carton**. Put a seed in the carton. Cover the seed with a little **soil**. Put some water on the soil. Wait a few weeks to see if a tree starts to grow.

Show what you know

Number your paper from 1 to 5. Write T for each sentence that is true. Write NT for each sentence that is not true.

1. All trees lose their leaves in cold weather.

2. Scientists tell a tree's age by counting its rings.

3. Many trees growing in one place make a forest.

4. Fire and insects can destroy forests.

5. If you plant a leaf of a tree and water it, a new tree will grow.

Think it through

Why are trees important to people?

4

Hidden animals

Find the animals

Some animals are hiding in these pictures. Do you see them? Point to the rabbit. Is the rabbit right side up or upside down? Find the deer. Is the deer looking at the rabbit? There are more animals in the pictures. Find them.

Now draw a picture. Hide something in it. Ask a friend to find the hidden surprise in your picture.

Now you see it

There is something unusual in the classroom. 1 What is it? Was it hard to find? Why?

Your teacher has hidden something in your classroom. You can see it from where you are sitting. Hold up your hand when you see it. Was it hard to find? Why?

Now it is your turn. Hide something in your classroom. Ask a classmate to find it. Was it hard to find? Why?

1

Now you don't

It is just like magic. These animals seem to disappear. 2, 3 Why? They look like the things around them.

The **moth** is the same color as the tree. Is the fish the same color as the things around it?

Find something to hide outside. Be sure it looks like the things around it. Ask a classmate to find what you hid.

2

3

Animal hunt

Walk around the school yard. Look for an animal that looks like things around it.

Look at **leaves** and **stems**. Look carefully into flowers. Look at the **bark** of trees. Look on the ground. Look into cracks in walls and sidewalks. Did you find an animal? What does it look like?

It is hard to see animals that look like things around them. 1, 2, 3 How do you think this helps the animals?

1

2

3

Show what you know

Number your paper from 1 to 3. Read each sentence. Find the right word in the Word Bag for each sentence. Write the word on your paper.

1. Sometimes animals seem to __________ .
2. A moth can be the same __________ as a tree.
3. A small animal can look like a __________ .

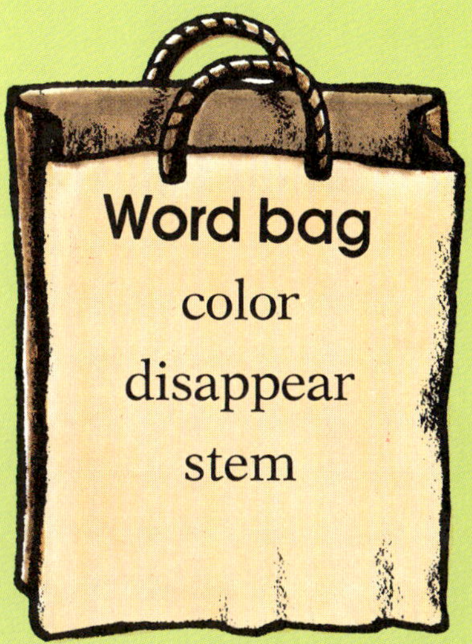

Think it through

Animals often look like the things around them. How does this help them escape danger?

5

Water animals

Plants and animals

These fish are in an **aquarium**. Some aquariums are jars. Others are glass bowls. All aquariums have water in them.

Water animals and water plants live in an aquarium. Fish and some kinds of **snails** are water animals. Find the fish and the snails in this aquarium.

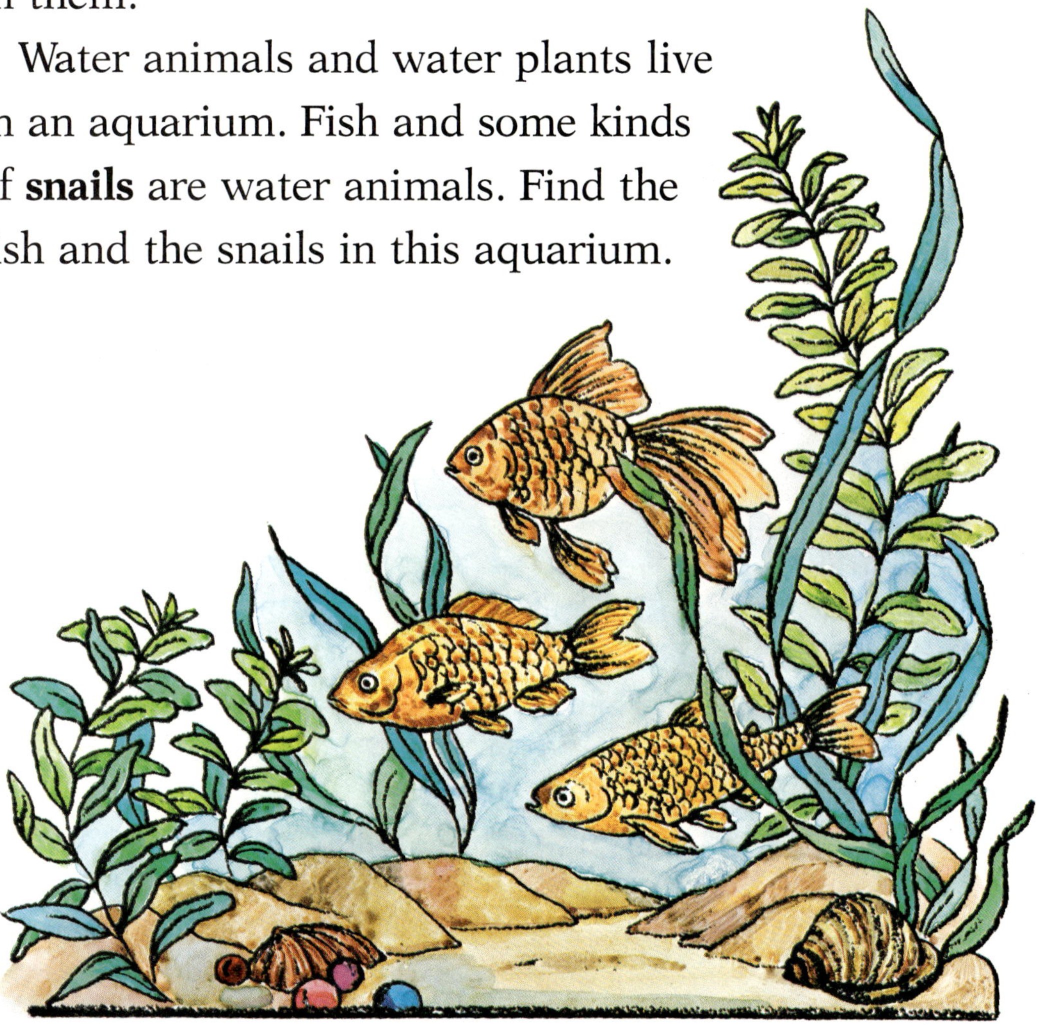

Activity

Make an aquarium

You can use a jar for an aquarium. 1 Put water and a water plant in the jar. Then, put in a fish and a snail. Watch your aquarium each day. What happens? Does your fish swim around much? Has your snail moved? **Record** what you see in your notebook. 2

1

2

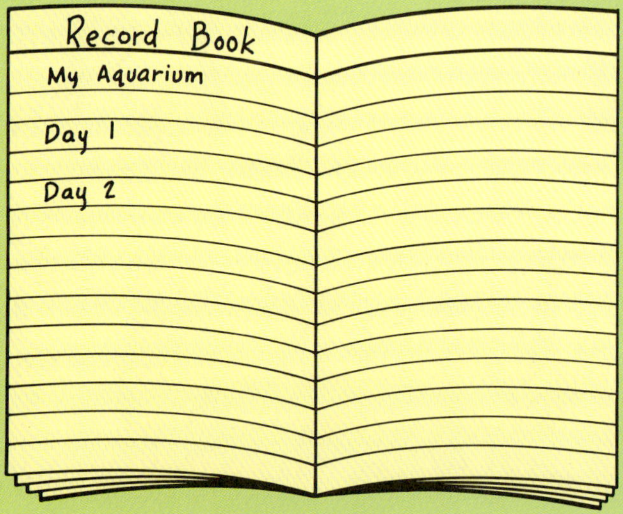

Snails

Each snail has a foot to help it move. Here are some snails. 3, 4, 5, 6 Can you find the foot on each one? Now look at your snail. Can you find its foot? Draw a picture of your snail's foot.

3

4

5

6

Fish

Fish are water animals, too. They have **tails** and **fins** to help them move. They have **gills** to help them breathe. Here are some fish. 1, 2, 3 Does your fish look like any of these? Can you see light through a part of your fish? Draw a picture of your fish.

1

2

3

A guppy

A **guppy** is a kind of small fish. Draw a picture of a guppy. First, draw the body. 4 Then, draw the tail fin and the eye. 5 Next, draw the other fins and the gills. 6

Look at your fish with a hand lens. 7 Do you think that your fish is a guppy?

4

5

6

7

Life in an aquarium

Here are some plants that live in water. 1 Is your water plant like one of these? Draw your water plant.

Be sure to put a little piece of fish food in your aquarium each day. Draw a picture of the way your fish eats.

Did you look at your snail's mouth? What other things do you see on your snail?

1

Show what you know

Number your paper from 1 to 5. Write T for each sentence that is true. Write NT for each sentence that is not true.

1. An aquarium has water in it.
2. Fish are animals.
3. Snails cannot live in water.
4. Every snail has a foot.
5. A guppy is a small dog.

Think it through

What can you learn from watching an aquarium?

6

Night animals

Seeing at night

Some animals like to sleep during the day when it is light. 1 And they like to wake up and go hunting at night. What night animals do you know of?

The eyes of most night animals have an extra part. This special part is in the back of the eye. It helps the animals see better in **dim** light. At night, the eyes of these animals seem to **glow**.

1

Owls

Owls are night animals. They can see in the daytime. But they see even better at night. Their big eyes let in lots of light.

Owls cannot move their eyes. They turn their heads to watch moving things. 1 Owls belong to the bird family. They fly quickly and quietly and make almost no noise.

1

Owls can be different sizes. The **elf owl** is only about 15 centimeters (about 6 inches) long. The great **gray owl** is nearly five times bigger.

There are many different kinds of owls. **2, 3**

2

3

Porcupines

Porcupines are night animals, too. They leave their tree-hole homes at night to look for food.

What is a porcupine like? Look at the picture. 1 The stiff **quills** on the porcupine's back and tail keep other animals away. The quills are very sharp. There are tiny hooks on the tips. They hook into and hurt anything they touch.

1

Hamsters

Hamsters are night animals. They like to sleep when it is light and wake up at night. 2

Golden hamsters make good pets. They are clean and gentle. And they are easy to care for. Hamsters live for only 2 or 3 years.

2

Watch a hamster. How? Make the animal think it is night. Bring a hamster to your class. Watch it sleep. Then, cover its cage with a dark cloth. 1 Be very quiet. The hamster may think it is night. Then, peek in, and watch the hamster. Is the hamster running around in its cage?

1

Show what you know

Number your paper from 1 to 5. Write T for each sentence that is true. Write NT for each sentence that is not true.

1. Night animals usually sleep during the day.
2. Night animals see best at night.
3. Owls can move their eyes.
4. Porcupines have quills in their tails.
5. Hamsters do not make good pets.

Think it through

Sometimes animals' names are used to describe people's **habits.** Habits are the things people do again and again. Why might someone be called a night owl?

7

Pandas

What are pandas?

Many people think **pandas** are bears. Why? Pandas stand up on their back legs as bears do. Pandas have short tails as bears do. And pandas climb trees the way bears do. But pandas are not bears.

So, what are pandas? Many scientists think pandas are part of the **raccoon** family. But some scientists say pandas are not bears **or** raccoons. These scientists say pandas are just pandas. What do you think pandas are? Why?

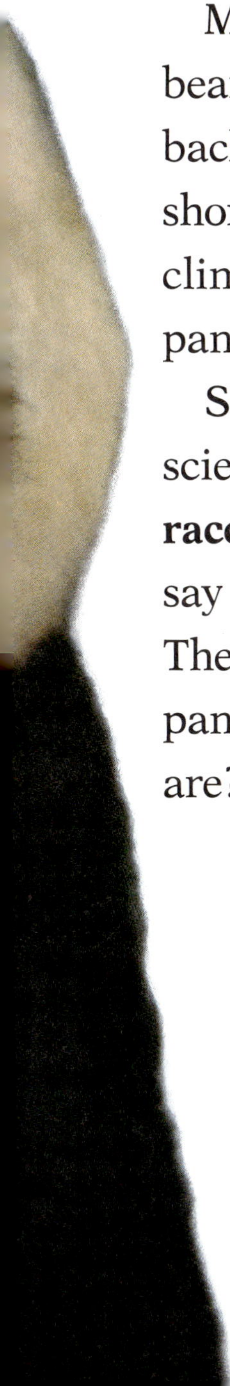

Two kinds of pandas

Most people know about **giant pandas**. These pandas are chubby black and white pandas. They are almost 1½ meters (about 5 feet) tall. And they weigh about 112 kilograms (about 250 pounds).

But did you know that there is another kind of panda? This is the **lesser panda**—sometimes called the red panda. 1 Lesser pandas are very small. Their bodies are just about ½ meter (about 2 feet) long. But they have very long tails. A red panda's tail may be 45 centimeters (about 18 inches) long.

1

Where do pandas live?

You may see a few giant pandas in zoos. But most pandas live in China. The pandas live in the mountains where there are lots of **bamboo** trees.

Pandas like to sleep in trees. They climb high up in trees to be safe from other animals. 2 The trees in the bamboo forest are very important to pandas.

2

What do pandas eat?

Bamboo is the favorite food of all pandas. 1 Pandas have big teeth and strong **jaws**. And they have one extra bone on each front **paw**. It is almost like having another finger. This helps pandas grab and chew the hard bamboo.

Sometimes, giant pandas will also eat small animals or leaves and roots of other plants. Lesser pandas sometimes eat insects, fish, and even birds. But, most of all, pandas like eating bamboo.

1

Show what you know

Number your paper from 1 to 4. Read each sentence. Find the right word in the Word Bag for each sentence. Write the word on your paper.

1. The black and white panda is the __________ panda.
2. The red panda has a very long __________ .
3. Pandas come from __________ .
4. Pandas usually eat __________ .

Think it through

How do you think the extra bone in a panda's paw helps the animal eat bamboo?

8

Water

Clean and dirty water

The water in Beaver **Brook** is very clean and **clear**. 1 There are large and small rocks under the water. These rocks are very clean. Fish swim in the water.

After a hard rain, leaves and twigs often float down Beaver Brook. They come from trees along the shore. The water in Beaver Brook does not get dirty. It does not get **oily**.

1

The water in Raccoon **River** is not clean and clear. You cannot see the muddy bottom of Raccoon River. The river is too dirty. Find Raccoon River on the map. 1

Here is a picture of the water in Raccoon River. 2 Where do you think the water in the river comes from? How does the water get so dirty? Do you think many people might live along Raccoon River?

1

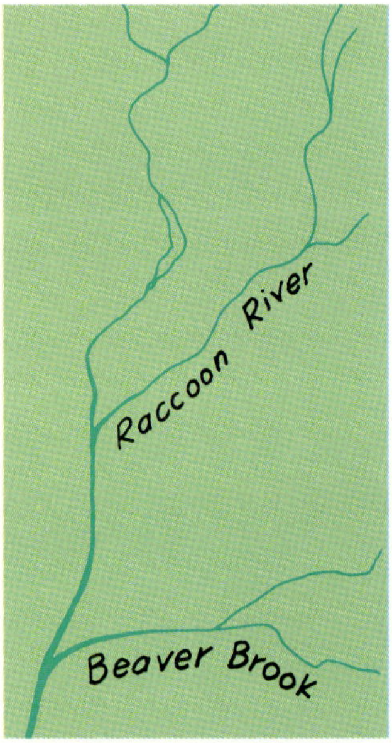

2

People make water dirty in many ways. Washing cars, streets, and dishes makes clean water dirty. In what other ways do people make water dirty? This is water from Beaver Brook and Raccoon River. 3 Which **waterway** has rocks on the bottom? Which has mud on the bottom?

3

Mixing water

Here is the place where Beaver Brook and Raccoon River meet. 1

What do you think the water looks like after the two waterways meet? The dirty water mixes with the clean water. The clean water becomes dirty. Are there some things in Raccoon River that you cannot see? What are they?

1

Activity

What happens to water?

Make some muddy water like the water from Raccoon River. Put some mud, salt, and dead leaves into a cup. 2 Fill the cup with clean water. Stir it. Is the water dirty now? What happened to the salt? Did it disappear?

Pour some of the dirty water into a cup. Put a lid on your cup. Keep the cup until your next class.

Is the water in your cup clear? 3 Where did the dirt go? Use a spoon to take some water from the top. 4 Is this water clear? Put the spoonful of water in the lid of your cup. 5 Save the lid until tomorrow.

2

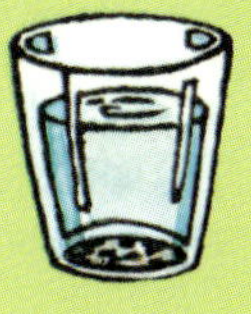
3

4

5

Cleaning water

Water is cleaned in many ways. Water is cleaned as it rushes over rocks on its way downstream. 1 Waterfalls help to clean water, too.

Some things, such as leaves and twigs, are easy to take out of water. Other things are hard to clean away. Did you get the salt out? Was there salt on the lid you saved?

1

Water can get very dirty and oily. It can get **polluted**. Polluted water is unsafe for drinking. It is unsafe for swimming and fishing. The water must be made clean before people can use it. 2

2

Where does the clean water in your classroom come from? 1 Do you think this water was ever dirty? How could it have been cleaned?

1

Show what you know

Number your paper from 1 to 4. Read each sentence. Find the right word in the Word Bag for each sentence. Write the word on your paper.

1. Water is made dirty in many ways by ________ .
2. Dirty water makes ________ water dirty.
3. Water is cleaned as it rushes over ________ .
4. Leaves and twigs are ________ to take out of water.

Think it through

What can people do to help keep water from getting dirty?

9

You grow up

Health and Safety

What can grow and change?

All living things grow and change. Tiny trees grow taller and taller. Tiny kittens become big cats. And human babies become grown-ups.

Do all of the things in these pictures grow and change? Which things are not living things?

Have you changed?

Have you had a birthday this year? Are you taller than you were last year? Do you **weigh** more than you did last year? Do you have more teeth than you did? Do you have fewer teeth? Have you changed?

What other things have changed about you in the past year? What other living things in your home have changed in the past year? How have they changed?

Activity

How much have you changed?

Make a chart to see how much you have changed. Ask your parents how much you weighed when you were born. Find out how tall you were. Write the **information** on your chart.

Find out how tall you are now. 1 And find out what you weigh. Write this information on your chart.

1 

Watch it grow

You are a living thing, and you grow and change. Plants are living things, too. So plants change and grow. You can watch a plant grow.

Start with a tiny seed. Plant the seed. Give it water, air, and sunlight. Keep a **record**. 1 Write down how the plant grows. Look at your plant every day. How do you and your plant change in the same ways?

1

Show what you know

Number your paper from 1 to 5. Write T for each sentence that is true. Write NT for each sentence that is not true.

1. Everything in the world grows and changes.
2. Dishes are living things.
3. You are probably taller than you were last year.
4. You have not changed much since you were a baby.
5. Plants are living things.

Think it through

How do you think you will change before you are as old as your parents?

10

DR. E. JOSEPH
FAMILY DENTISTRY & ORTHODONTICS
J. HALLMARK INC.
REAL ESTATE
ENTR REAR PARKING LOT-SUITE 7L
QUEENS CONSULTATION & COUNSELING SERVICE
DR. JOSEPH J. WITT
PODIATRIST
FOOT SPECIALIST

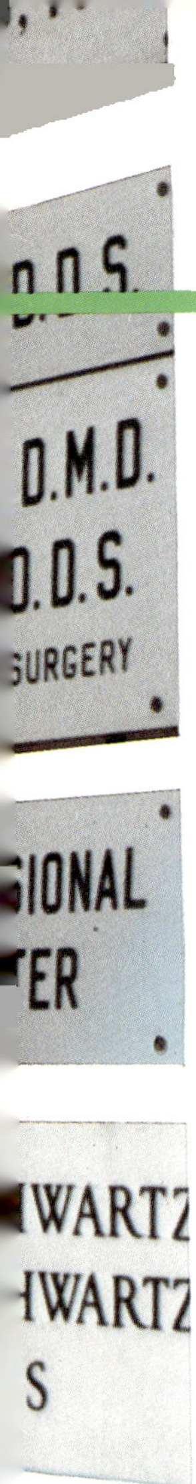

Stay healthy

What you can do

Doctors take care of people who are sick. And doctors help people learn how to stay **healthy**.

How can you stay healthy? Eat good food. Get plenty of **exercise**. Keep clean. Brush your teeth. Get lots of rest. And get **checkups**.

Who gives checkups? Doctors do. There are many kinds of doctors. Each one has a special job to do.

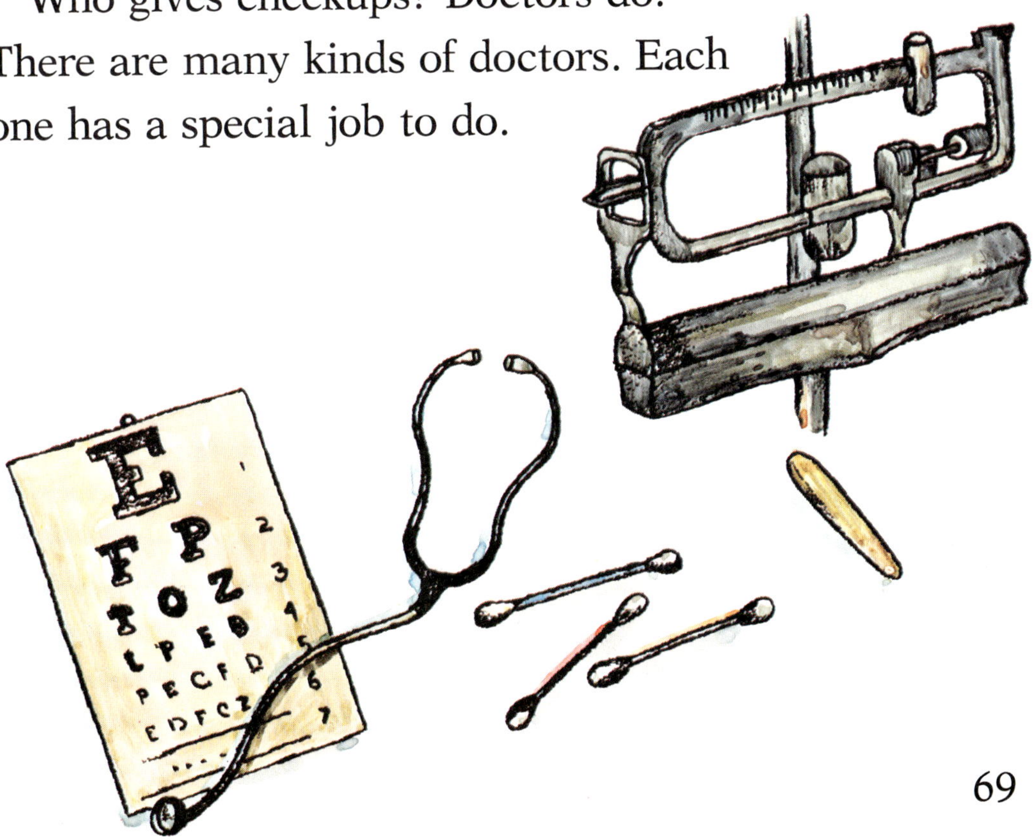

Doctors check children

Some doctors work only with children. These doctors help children grow up healthy. Patients get checkups once a year from their doctors.

What happens during a checkup? Your **weight** and **height** are checked to see how you are growing. 1

1

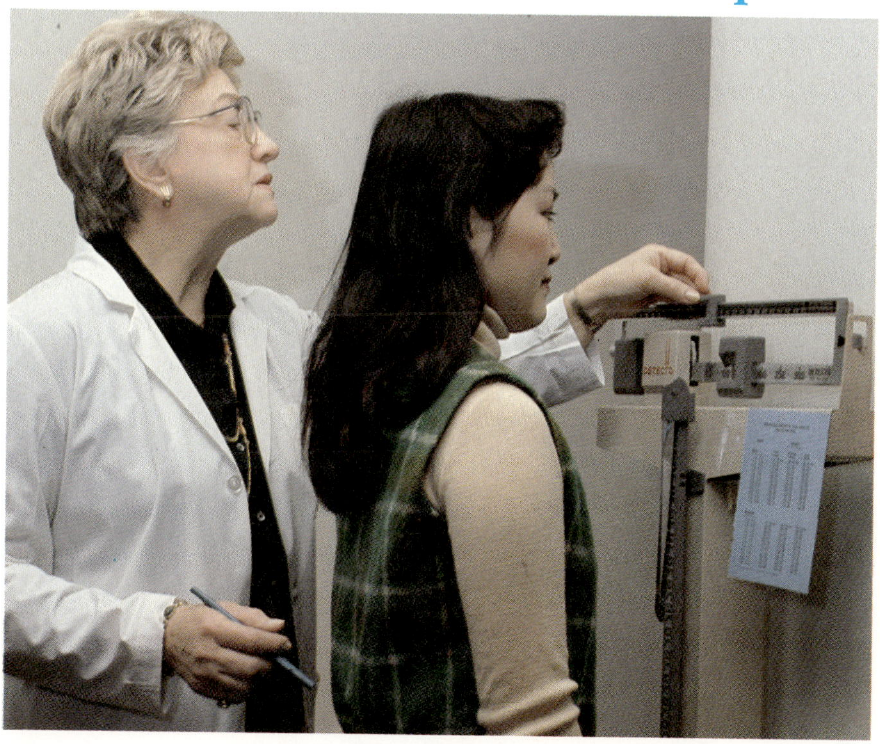

Your **temperature** is checked. The doctor listens to your heart and looks in your ears. Your throat is checked as you say, "Ah-h!"

The doctor also checks your **blood pressure**. 2 And you may get a shot that helps you stay healthy.

2

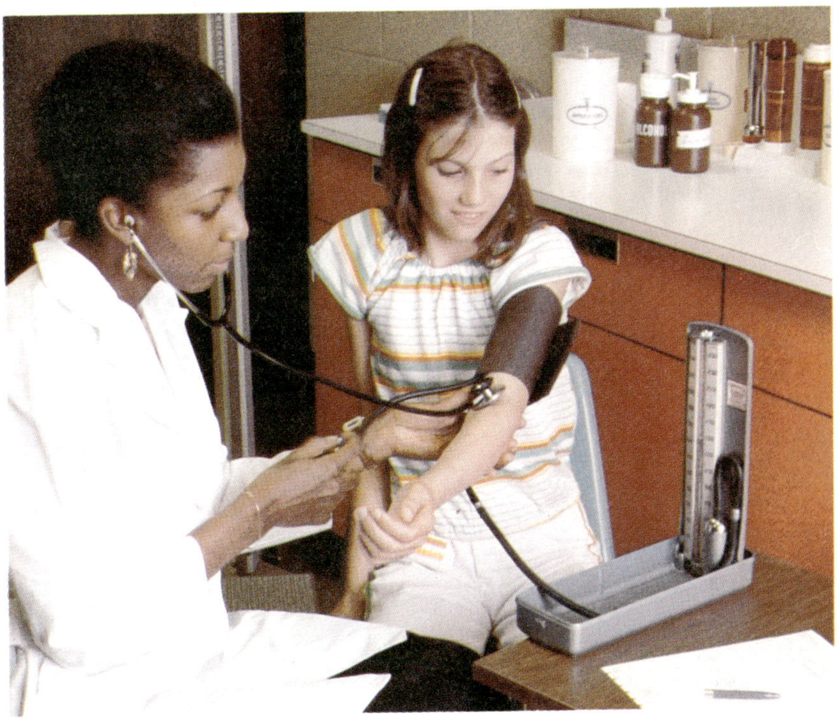

Doctors check teeth

Some doctors work only on teeth. These doctors are called **dentists**.

When you go for a checkup at the dentist's office, you sit in a big chair. The dentist in the picture says, "Open wide!" 1 He turns on a bright light and looks into your mouth. A little mirror helps him see all your teeth.

1

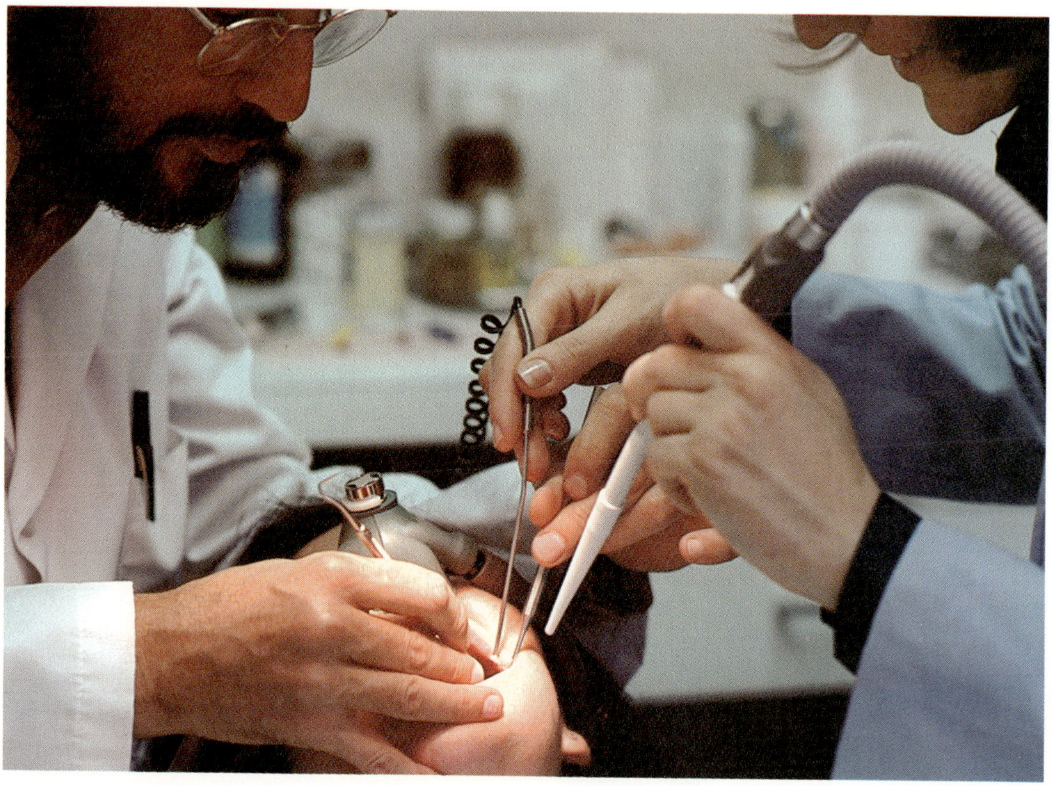

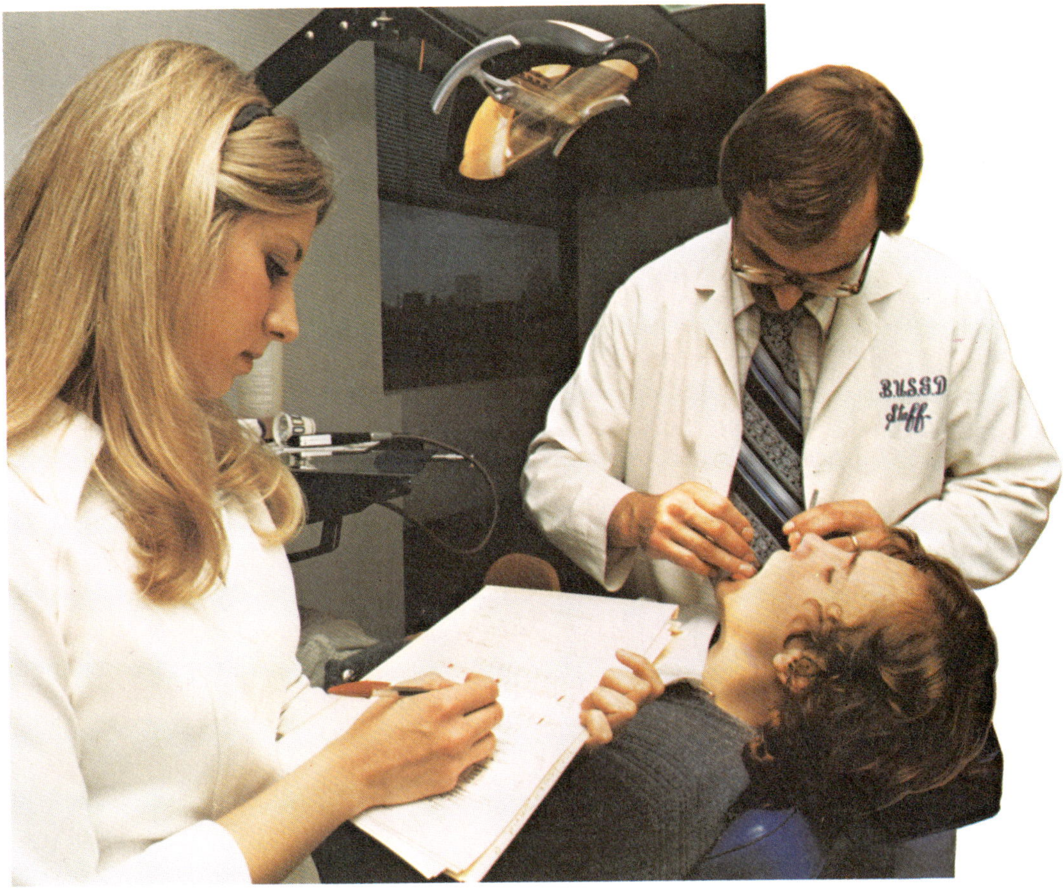

2

The dentist may find a little hole in a tooth. 2 That hole is called a **cavity**. How can the dentist fix it? First, it must be cleaned out. A machine is used to clean out the cavity. Then, the dentist puts a filling in the hole.

The dentist or a helper may clean your teeth, too. The dentist may take **X rays** of your teeth. Why? X rays show how well your teeth are growing. The dentist will tell you, "Be sure to brush your teeth. I'll see you in 6 months."

Doctors check eyes

Some doctors take care of eyes. These doctors use eye charts. They use lights to look into each eye. And they use special machines to check **vision**. 1, 2 Sometimes doctors put medicine in eyes, too. Eye doctors tell people how to take care of their eyes.

1

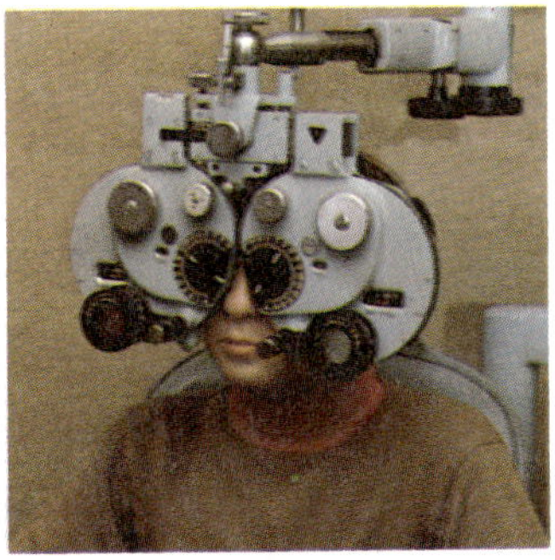

2

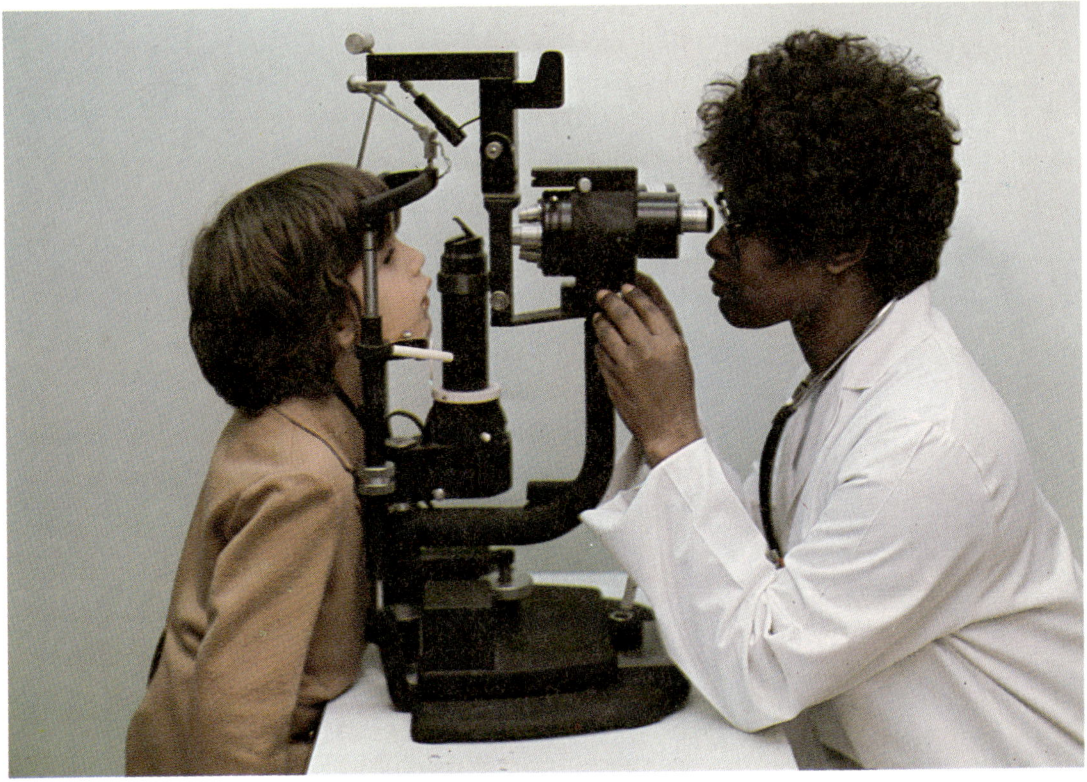

Sometimes doctors find that people need glasses to help them see better. The doctors order glasses. Each pair of glasses is special because it is made for just one person.

Show what you know

Number your paper from 1 to 5. Read each sentence. Write the word that best completes the sentence. Use the words from the Word Bag.

1. Checkups tell you how __________ you are.
2. __________ check teeth.
3. A __________ is a hole.
4. Some doctors use special __________ .
5. Eye doctors order __________ for people.

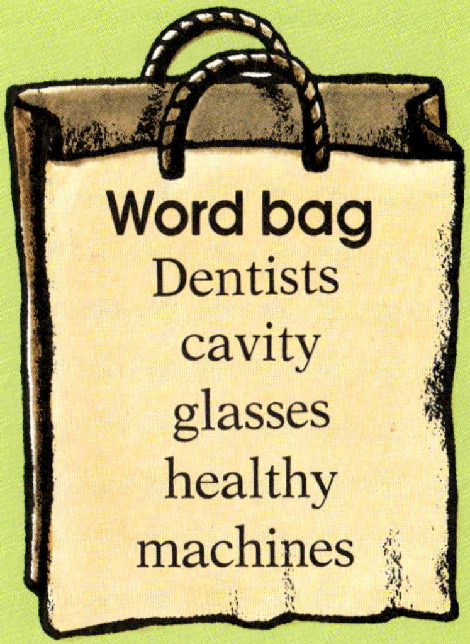

Think it through

Why do you think it is important for people to have checkups?

11

Water safety

Spending time in the water

Fish and ducks live in water. So do lots of other animals. Many people spend time in, on, or near water, too. Do you?

Every year, many people are hurt in water **accidents**. Why? Some people do not follow safety rules.

Do you know any **water safety** rules? On the next five pages, you will find some important water safety rules to learn.

1

Swimming safely

Can you **swim**? Millions of people do. Swimming is good **exercise**. 1 But it can be harmful if you do not follow the rules for safe swimming.

Here are six important rules of swimming safely. Can you think of any more rules of safe swimming?

Swimming Safety Rules

1. Learn to swim.
2. Swim only when a lifeguard is watching.
3. Always swim with someone.
4. Do not swim right after eating a big meal.
5. Swim only in a safe place.
6. Do not call for help unless you need it.

Some people swim in pools. Others swim in lakes and oceans.

Look at these two safety signs. 1, 2 How are the rules alike? How are the rules different?

1

2

Boating safely

Many people like boats. 3 Rowboats, sailboats, kayaks, and canoes are boats. They are great fun. But they can be dangerous if you do not know how to use them.

Read the Boat Safety Rules on the next page. Can you think of more ways people can stay safe while boating?

3

Boat Safety Rules

1. Always wear a life jacket in a boat.
2. Stay seated in a boat.
3. Have someone hold the boat whenever you get in or out.
4. Always tell someone on land where you are going.

Show what you know

Number your paper from 1 to 5. Write T for each sentence that is true. Write NT for each sentence that is not true.

1. Swimming is good exercise.
2. You should never swim alone.
3. You can swim right after eating a big meal.
4. You should not play on diving boards.
5. You do not need to wear a life jacket in a boat if you are a good swimmer.

Think it through

Why do you think it is important for everyone to learn how to swim?

2

Earth and space science

12

Rocks

Granite

There are many different kinds of rock. 1 One kind is **granite**. It has black spots. It has white spots. Granite is made of lots of pieces. The pieces may be large or small. And the pieces may be many different colors.

Granite can be part of a mountain. Granite is a very hard rock. It is used for building. Look around. Do you see some ways in which granite is used?

1

Activity

Looking at granite

Ask your teacher for a rock. Does it look like granite?

Use your hand lens to find black pieces in the rock. 1 Are there white pieces in it? What other colors do you see? Draw a picture of what you see.

1

Limestone

Here is another kind of rock. 2 It is called **limestone**. Limestone does not have spots like granite. But there is something else in limestone. Can you see what it is? Limestone has **fossils** in it. 3 Fossils are the marks made by a plant or an animal. The plant or animal lived very long ago. A fossil may be a footprint of an animal. Or a fossil may be a **print** of a leaf.

2

3

Looking at limestone

Look at some limestone with your hand lens. Do you see something in the limestone? What does it look like? Draw a picture of what you see.

Find a rock outside. 1, 2 Are there small pieces in the rock? Is the rock granite? Is it limestone?

1

2

Show what you know

Number your paper from 1 to 3. Read each sentence. Write the word that best completes the sentence. Use words from the Word Bag.

1. Granite is made of lots of little __________ .

2. Some __________ are granite.

3. __________ is a rock that has fossils in it.

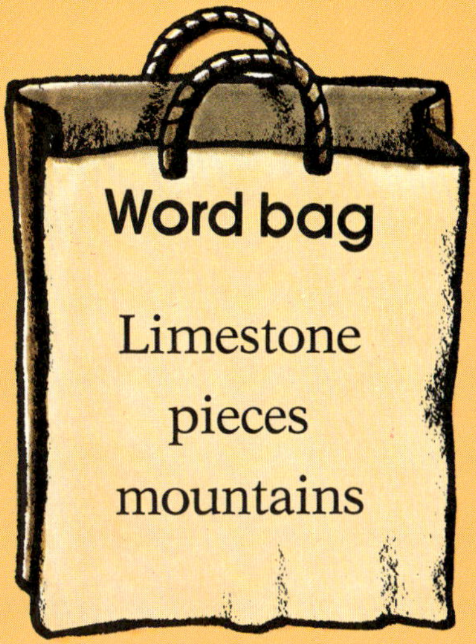

Think it through

How do you think the fossils got into the limestone?

13

Life long ago

Brontosaurus

Brontosaurus lived in Wyoming long ago. It was one of the largest **reptiles** that ever lived. It was as big as five elephants. When Brontosaurus walked, the ground shook. Its name means thunder **lizard.** Brontosaurus ate only plants. Most of the time, Brontosaurus rested and ate.

Diplodocus

Diplodocus was another reptile that ate plants. 1 It lived in Utah long ago. Diplodocus was probably the longest land animal that ever lived. It grew to nearly 30 meters (about 100 feet) long. Diplodocus stayed in the water most of the time. There, it was safe from enemies.

1

2

Tyrannosaurus

Tyrannosaurus was an enemy of all animals long ago. 2 Its name means **tyrant** lizard. Tyrannosaurus killed and ate all kinds of animals. Tyrannosaurus stood on two large legs. It also had two smaller arms and a big tail.

Tyrannosaurus stood 6 meters (about 20 feet) tall. One of its feet was bigger than the top of your desk. When Tyrannosaurus walked, its footprints were more than 3 meters (about 10 feet) apart.

The big reptiles died long ago. Only some of their bones, teeth, and footprints are left. 1

1

Ancient ancestors

Some animals of long ago were **ancestors** of animals that live now. 2 Today's lizards and snakes probably are descended from snakes and lizards of long ago.

What animal of today is like **Protorosaurus?** 3 Does an alligator look like Protorosaurus? 4

2

3

4

Many insects are descended from insects that lived long ago. An ancient **dragonfly's** wingspan was more than 60 centimeters (about 2 feet). 1 The dragonfly of today has a wingspan of only 10 or 12 centimeters (about 4 or 5 inches). 2

Today's Dragonfly

2

1

Ancient Dragonfly

Some of the earliest birds had claws on their wings! They used the claws to climb trees. Then they sailed off the trees. These birds probably did not fly very well.

Show what you know

Number your paper from 1 to 5. Write T for each sentence that is true. Write NT for each sentence that is not true.

1. Brontosaurus lived in New York.
2. Brontosaurus ate plants.
3. Diplodocus was very small.
4. Tyrannosaurus ate other animals.
5. Tyrannosaurus left footprints.

Think it through

How do scientists find out about animals of long ago?

14

Clouds

What are clouds?

Did you know there are always many drops of water in the air? You cannot always see them. But they are there. And there are tiny bits of dust in the air. But you cannot always see them, either.

Sometimes the bits of dust and the drops of water stick together. Then you can see the water and dust. Why? Because together they make a **cloud.**

Activity

Make a cloud

You can find out how a cloud is formed. Use a clean bottle. Put some hot water into the bottle. 1
Now place an ice cube on top of the bottle. 2 What do you see in the bottle?

The hot water in the bottle makes some of the air warm. The warm air floats up. It hits the ice cube. And a cloud is formed.

1

2

Some clouds are white

Have you seen fluffy white clouds? Did you find shapes in the clouds? Sometimes clouds look like sheep or birds. What shape do you see in these clouds? 3

3

Maybe you have seen other white clouds. These clouds are shaped like feathers. 1 They are very high in the sky.

1

When you see white clouds, it is usually a nice, sunny day. The clouds look white when the sun shines on them.

Some clouds are dark

Have you ever seen this kind of cloud? 2 What kind of day was it?

Do you know what made the cloud dark? Remember, every cloud is made of tiny drops of water. Sometimes the drops get bigger and bigger. When the drops get very heavy, the cloud looks dark. Rain begins to fall. Why? The big water drops are too heavy to float high in the air.

2

The sun cannot shine through very dark clouds. See for yourself. Put a flashlight behind a piece of white paper. 1 Does the light show through? Now try the same thing with a piece of black paper. 2 What happens?

Show what you know

Number your paper from 1 to 5. Read each sentence below. Write T on your paper for each sentence that is true. Write NT for each sentence that is not true.

1. A cloud is made from bits of dust and drops of water.
2. Warm air goes up.
3. White clouds are rain clouds.
4. Raindrops fall because they are too heavy to float.
5. Light shines through dark clouds.

Think it through

What do you think happens when raindrops start to fall on a very, very cold day?

15

Weather

Reporting the weather

Will it be sunny and warm today? Or will it be cold and rainy? Will there be a windstorm? These are some things that **weather reporters** tell us.

Weather reporters use many different **instruments** to check the weather. They need to know what is happening to the three big weather makers.

The sun is a weather maker

The sun is hot. In the summer, the sun is right above our part of Earth most of the time. 1 The hours of daylight are long. The sun keeps us warm in summer. In fact, sometimes we get too hot.

1

But, in winter, the sun is right over other parts of Earth. Our hours of light are shorter, and our nights are long. So the sun does not warm the air as much in winter. We feel cold.

Do you think it is cold or warm here? **2** The sun is shining, but it is not warm here. Why?

2

Water is a weather maker

You know that clouds are made of water drops. Sometimes the water in the clouds falls as rain. Sometimes a cloud comes close to the ground as **fog**. 1 Or maybe it snows. Water makes weather in many ways.

1

Activity

Water gets into air

How does water get into the air? Let's find out. Use two jars. Pour a cup of water into each jar. 2 Put a lid on one jar. Do not cover the other jar. Watch the jars for a few days. What happened? 3 Does one jar have less water? Where did the water go?

2

3

Wind is a weather maker

Wind is moving air. Sometimes air moves slowly. And sometimes air moves fast. If the weather is good, the wind may be light. But strong winds can make trouble. 1

You can find out if the wind is slow or fast. Make a **wind machine** like this one. 2 The four paper cups catch the air. The harder the wind blows, the faster the cups turn. Is the wind blowing fast or slowly in this picture? 3

1

2

3

Show what you know

Number your paper from 1 to 5. Read each sentence. Write the word that best completes the sentence. Use words from the Word Bag.

1. The sun is above our part of Earth in ________ .
2. The days are short in ________ .
3. A cloud close to the ground is ________ .
4. ________ is moving air.
5. ________ winds make trouble!

Think it through

What happens to wet clothes if you hang them outside? Why?

16

Up in the sky

The sun

Do you see that bright light in the sky? It is the **sun**. The sun is a big ball of hot **gases**. It is much larger than Earth. But it looks smaller because it is millions of miles away.

The sun gives Earth heat and light. Without the sun, plants and animals could not live.

Earth moves around the sun. Earth turns as it moves. The part of Earth turned toward the sun has daylight. And the part turned away from the sun has night.

The moon

You may see the **moon** in the night sky. But the moon does not always look the same. Sometimes the moon will look like a round ball. 1 At other times, it will look like only a part of a ball. 2, 3 The moon seems to change shape.

1

2

3

You can make a chart about the moon. 4 Look for the moon every night at the same time. Did you see the moon every night? Was the moon in the same place at the same time each night? Did the shape of the moon seem to change?

4

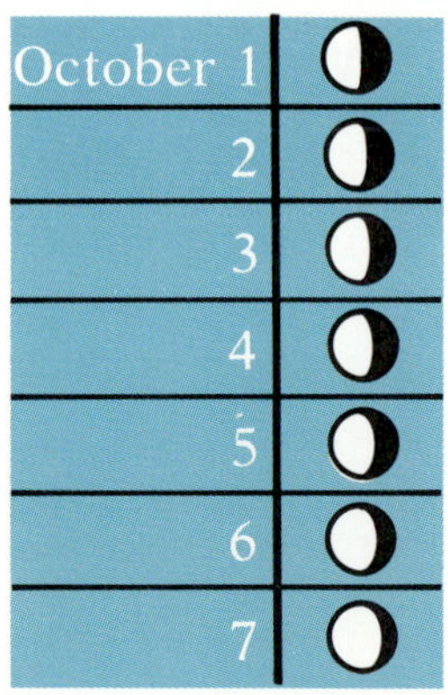

The planets

Planets rise and set. Planets are bright, too. **Venus** is a planet. You may see Venus just after sunset or just before sunrise. 5

Scientists are learning more about planets all the time. Watch the newspapers for stories and pictures about planets. This is a picture of the planet Saturn. 6

5

6

The stars

Like the sun, **stars** give off light. So stars twinkle and shine in the night sky.

Look north at night. You may see one or more of these three groups of stars. You may see **Cassiopeia**. 1 It looks like a broken chair. You may see the **Big Dipper**. 2 It is sometimes called the Great Bear.

1

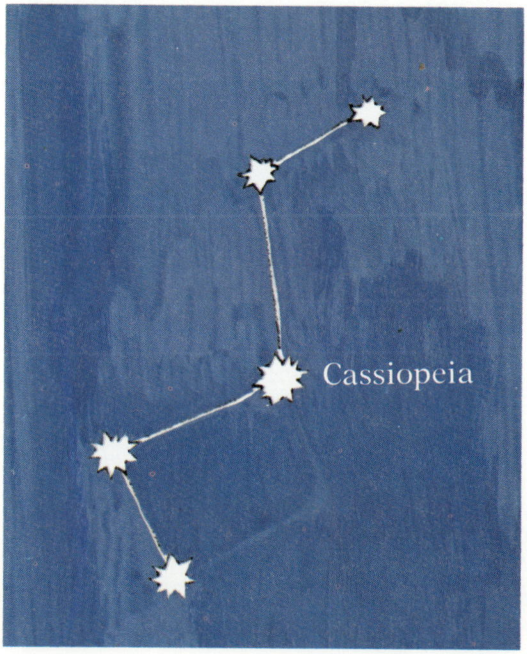

2

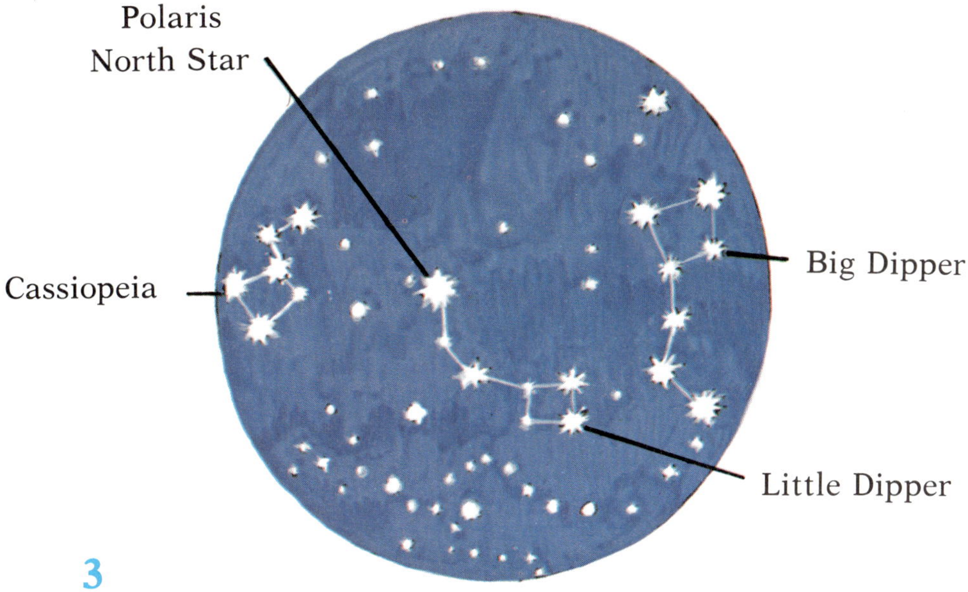

3

You may see the **Little Dipper**. 3 It is sometimes called the Little Bear. Look at the very bright star at the end of the handle. That is **Polaris**, the North Star.

You can use Polaris to help you tell direction. Just remember that when you face Polaris, you are facing north.

Look for Polaris tonight. Look halfway between the Big Dipper and Cassiopeia. Is the Big Dipper east or west of Polaris?

Spaceships and satellites

People send things to the sky. **Spaceships** are sent to the sky by people. The spaceships are pushed up by rockets. 1

Some spaceships go to the moon. Other spaceships go far into space to other places. Machines on the spaceships send messages and pictures back to Earth. The messages and pictures help people learn more about space.

People send **satellites** into space, too. The satellites go around and around Earth. Some satellites take pictures. Maybe you have seen weather-satellite pictures. These pictures show how clouds and storms move. Other satellites help send TV shows around the world. Have you ever seen a TV show sent by satellite? Did the show come from another country?

1

Show what you know

Number your paper from 1 to 5. Read each sentence. Write the word that best completes the sentence. Use the words from the Word Bag.

1. The sun is a ball of __________ .
2. Earth moves around the __________ .
3. Venus is a __________ .
4. The North Star is called __________ .
5. Some __________ take pictures.

Think it through

Suppose you were lost in the woods. How could Polaris help you find your way?

3

Physical science

17. Float and sink
18. Compasses
19. Lights
20. What is hot?
21. Push and pull

17

Float and sink

Can you float?

Can you **float** in water? Some animals can. Some people can. Some things float, and other things **sink**. Sometimes things that usually sink can float. And sometimes things that usually float will sink.

Helping clay float

Put a lump of clay in water. It sinks. Now put a wooden stick in water. What happens to the stick? What would happen if you put wooden sticks in the clay? Try it.

What other things might you use to float clay? Corks would work, but nails would not. Why? Try to float your clay with drinking straws. 1

1

The inside story

This looks like a ball of clay. But the ball floats. 2 It has floated for a whole week. How? Something is in the ball of clay. What do you think it is? It could be a cork or a piece of wood. It could even be air. Cork, wood, and air help things float. Do you think a rock could help something float?

2

Put something in a ball of clay. Does the clay float? Try many different things. Make a list of the things that help the clay ball float.

Activity

Making things float

Put two cans like these in water. 1 One sinks and one floats. Why? What is different about the can on the left?

1

Make a little clay can. Put it in water. Now make a clay cap for the can. Put it in the water, too. Do they float? Put the cap on the clay can. What happens now? Why?

Boats and blocks

Flatten some clay. Make it look like a turtle shell. Put your clay turtle shell in water. Does it float?

Turn your clay turtle shell upside down. What happens?

Now make a clay boat like this one. 2 Put it into the water to find out if it floats.

2

Put a wooden block in your clay boat. Add more blocks until the boat begins to sink. 1 How many wooden blocks did you use? What happens to the wooden blocks when the clay boat sinks?

Shape a piece of **aluminum foil** into a boat. Put blocks into the boat until it sinks. How many blocks did you use?

Which makes a better boat, clay or foil? Why?

1

Show what you know

Number your paper from 1 to 3. Read each sentence below. Choose the word that best completes the sentence. Find the word in the Word Bag.

1. Wood __________ .
2. Sometimes a thing floats because of something __________ it.
3. If a boat has too many things in it, the boat will __________ .

Think it through

What could you use to cover the outside of a ball of clay to make it float?

18

Compasses

Direction finders of long ago

Long ago, people found directions in different ways. Sometimes people used the stars, the sun, or **moss** on trees to help them find directions.

Green moss grows on the **north** side of trees. Can you see the green moss in the picture of the tree? Then you know which way is north. Which way is **south**?

This is a picture of farmland. 1 It is not nighttime. The sky is too cloudy to show the sun. How could you find directions here? Would a **compass** help? To learn how a compass works, you must know about **magnets**.

1

Magnets help

Magnets can pick up some things. 2 Get a magnet. Try to pick up a pencil with the magnet. Try to pick up crayons and other things. Pick up a paper clip. Magnets pick up some metal things. They also stick to metal.

2

How does a magnet act with another magnet? Bring your magnet near another magnet. Do the magnets pull together? Now turn one of the magnets around. Do the magnets push apart?

Activity

A floating magnet

Use a pencil to make holes in a foam ball. Make four holes in a row. **1**

Push your disk magnet into the holes. Your ball and disk should look like this. **2**

Put your ball in a cup. Did the disk roll to the bottom? Pour a little water into the cup. Does the ball move?

Take the ball out of the cup, and push a toothpick through it. **3** The toothpick will be your pointer. Place the ball back in the cup. Watch the ball until it stops moving. Where is the pointer pointing? Now turn the cup. The ball will move and stop. Look at the pointer. It is pointing in the same direction as before.

1

2

3

A floating compass

Now put the cup and floating magnet on the table. The magnetic ball will turn slowly and then stop. 4 Why? The magnetic ball stopped when the pointer found north. Mark the letter N on your cup. It should be at the sharp end of the toothpick pointer. 5 Mark the letter S at the other end. 6 You have now made a compass. Walk around the classroom with your floating compass.

4

5

6

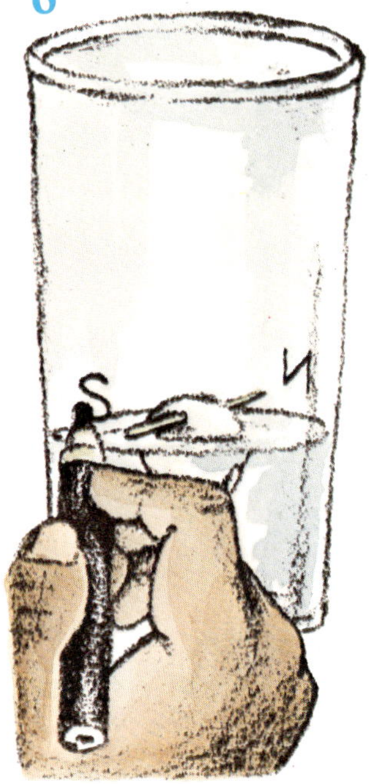

What could keep your floating compass from pointing north? Some metals attract magnets. The compasses worked near this wooden table. 1 But they will not work near a metal desk. Why?

1

Remember that magnets act on each other. They may bump together. They may push apart. A magnet will stop a compass from pointing north.

2

What will happen if you put two floating compasses into one dish? 2 Will they point north? Try it, and see.

Many people use a compass to find directions. Sailors use a compass to check their position at sea.

Now you know how to find north on farmland. Think of some other times you might use a compass.

Show what you know

Number your paper from 1 to 5. Read each sentence. Then write the word or words that belong in the sentence. Find the word in the Word Bag.

1. Tree moss, compasses, and __________ can be used to find directions.
2. A __________ does not pick up a crayon.
3. Two magnets can __________ and __________ each other.
4. A compass points __________ .
5. A magnet will stop a __________ from pointing north.

Think it through

In what other ways are magnets used?

19

Lights

All kinds of lights

Lights come in all shapes, sizes, and colors. There are lights for parking lots, playgrounds, streets, stores, planes, and trains.

Lights light up nearly everything, from walkways to watches. How are the lights used in these pictures?

Lights on

Light **bulbs** come in all shapes and sizes, too. 1 Some are as big as a balloon. Some are as small as a grain of wheat.

1

You can use **wire** and a **battery** to light a bulb. A bulb has two **contacts**. 2 Here are two ways to connect the contacts to the battery. 3, 4 Both ways work. Now look at these bulbs. 5, 6 Why won't these bulbs light?

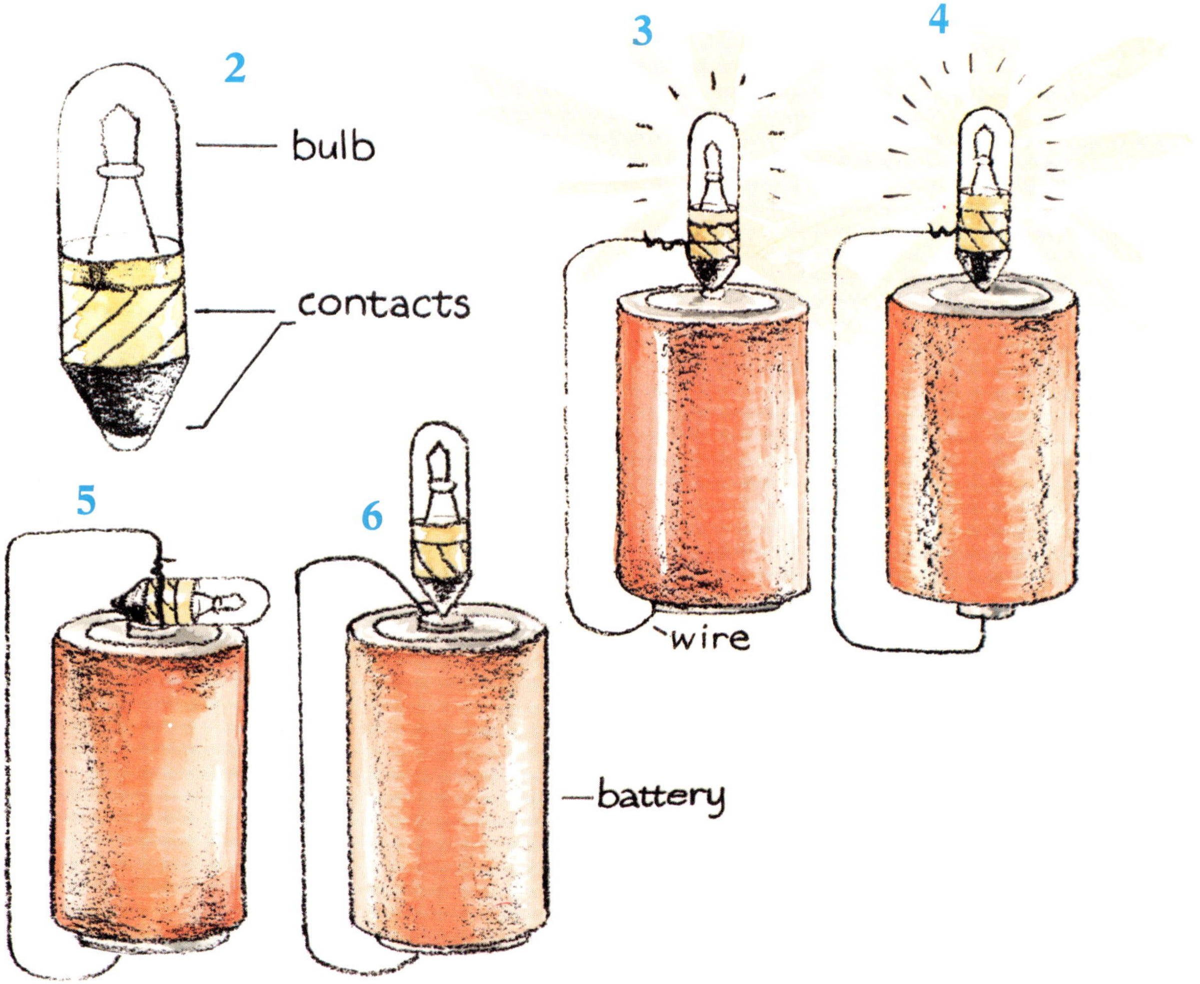

The bulbs light when there is a **circuit**. A circuit lets **electricity** flow. Electricity flows through the wire and through the bulb. 1 It goes from one end of the battery to the other.

1

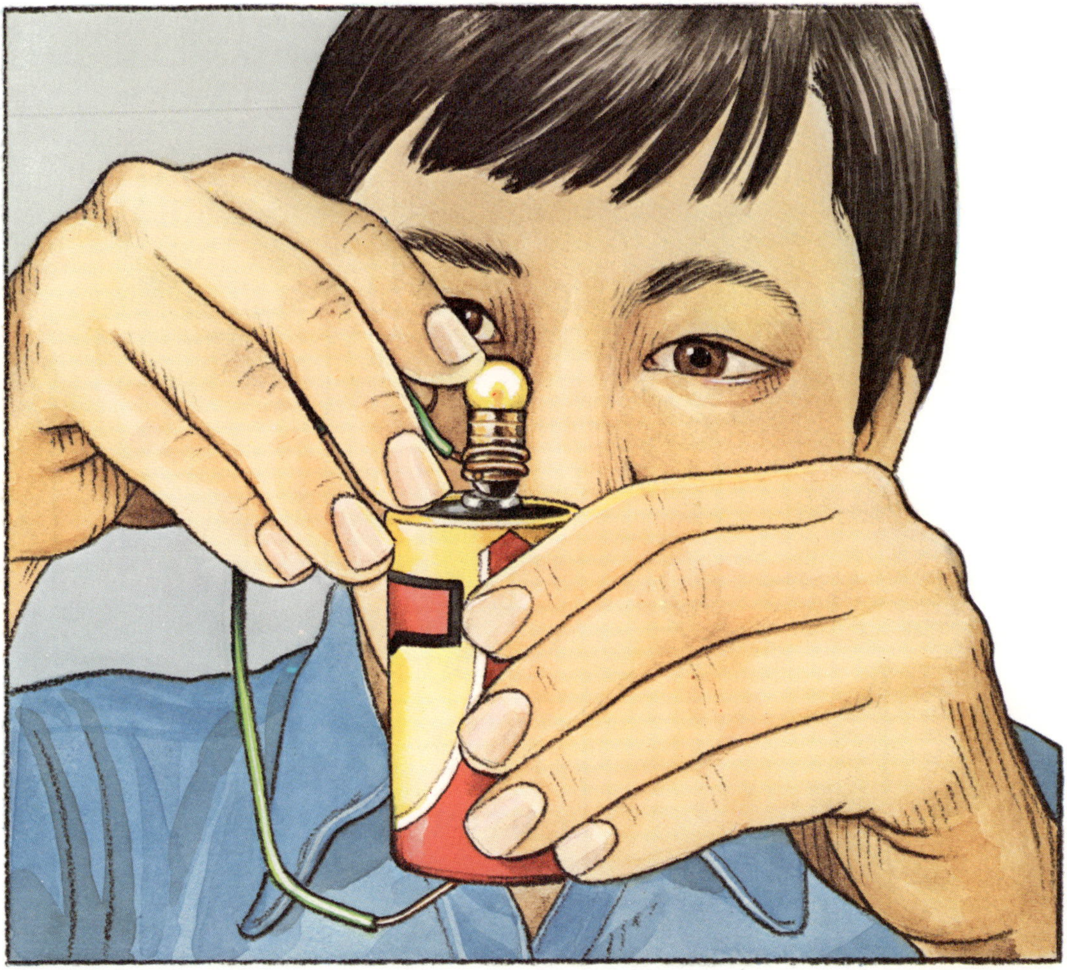

On and off

Should one light be on and one light off in this picture? 2 Look at the cup with the green spot. Only one wire is touching the battery. The circuit is not complete. The bulb under that cup cannot be on.

2

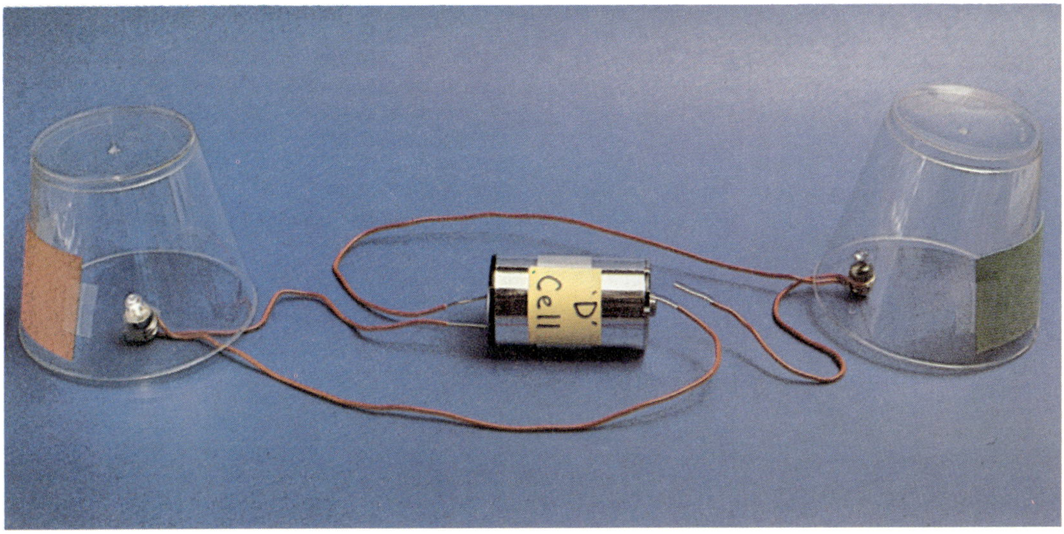

The two wires from the other cup touch both ends of the battery. The two contacts on the bulb are also connected to the battery. The bulb is on because a complete circuit has been made.

Try some different ways to connect a bulb to a battery. Try a paper clip and foil. 1 Do you think the bulb will light? Yes, it will. Nails and pins will make the light go on, too. But rubber bands will not. Which things on the table will complete a circuit? 2

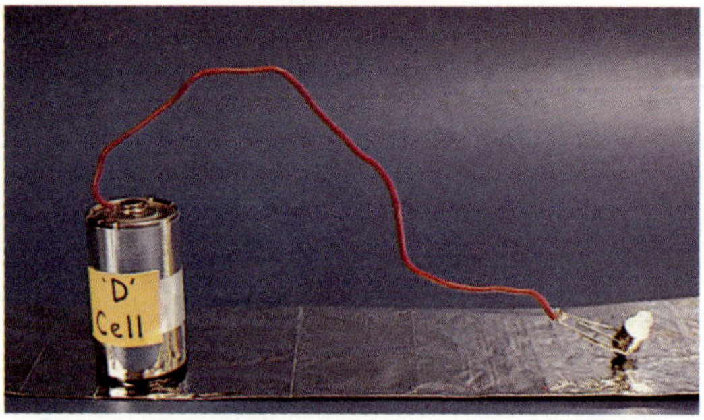

1

2

Activity

Lots of light

Try to light more than one bulb. First, put one bulb on each end of a battery. 3 Next, try two bulbs at the same end. 4 Do all the lights go on? Are some lights brighter than others? Why do you think this happens? Try it, and see.

3

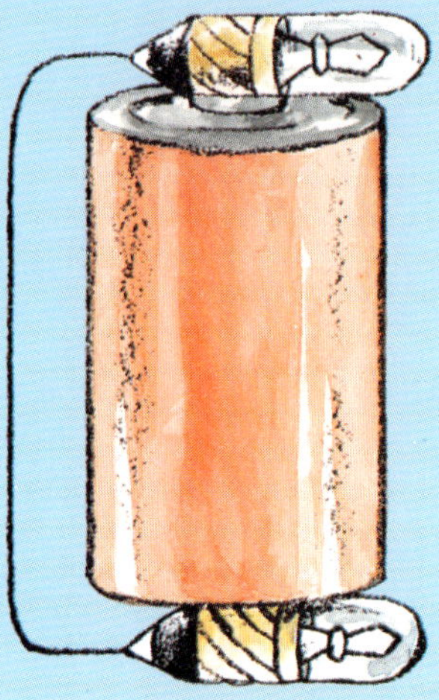

4

Try to light a bulb with two batteries. 1 It works. The light is very bright. Try to light five bulbs with one battery. 2 Which bulb do you think will be brightest? It will be the one closest to the battery.

REMEMBER! NEVER TRY TO USE ELECTRICITY FROM WALL OUTLETS OR LAMPS. THE ELECTRICITY IS TOO POWERFUL. IT CAN HURT YOU!

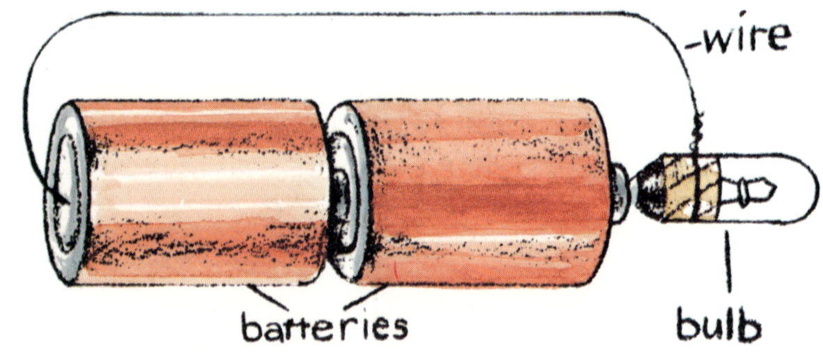

1

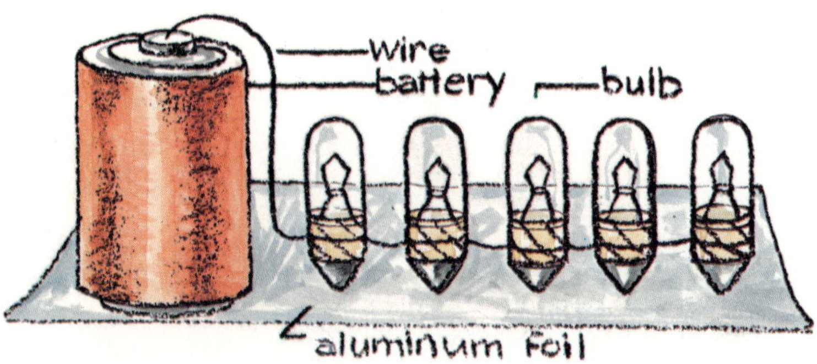

2

Show what you know

Number your paper from 1 to 5. Write T for each sentence that is true. Write NT for each sentence that is not true.

1. A circuit lets electricity flow.

2. You can get electricity from a battery.

3. A complete circuit makes a light go off.

4. Paper clips and aluminum foil can complete a circuit.

5. Rubber bands can complete a circuit.

Think it through

Do you think a battery will keep a bulb lit all night. Why?

20

What is hot?

Temperature

The boy wants to melt the butter. He holds it in his hand, but it does not melt quickly. So he decides to put the butter in a warmer place. He needs a warmer **temperature** to melt the butter. Temperature tells us how hot something is. And temperature tells us how cold something is.

How did the boy melt the butter? Where did he find a warmer temperature?

Room temperatures

Check the temperature around your classroom. 1 Check it on the floor and near the ceiling. Check the temperature near a window. Now check it near the door.

1

Make a list of your findings. 2 Is the temperature higher on the floor or near the ceiling? Is the temperature higher in the sun or in the shade?

2

Classroom Temperatures	
Place	**Temperature**
1. ceiling	68°
2. floor	63°
3. windowsill	65°
4.	
5.	

Where is the hottest spot in your classroom? Were you surprised by your findings?

Temperatures in bottles

A **terrarium** is a small place for plants and animals. 1 Read the temperature inside the terrarium.

An **aquarium** is a small place for water animals and water plants. 2 Read the temperature inside the aquarium.

The terrarium and the aquarium are in the same room. They are on the same table. Why do you think the temperatures are so different?

1

2

Activity

Warming up

Make a little aquarium by putting water into a cup. **3** Next, make a little terrarium by putting sand in another cup. **4** Put a thermometer in each cup. Are the temperatures the same?

Now, put a lid on each cup. Put both cups in the sun. Watch the thermometers for 5 minutes. Record the temperatures. Look again later in the day. Which cup is warmer?

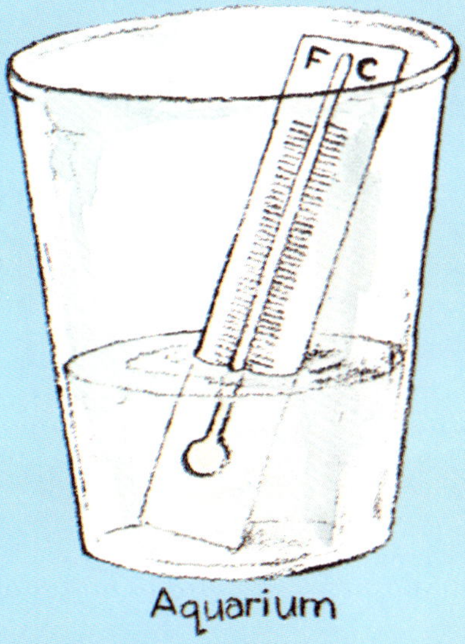

3

4

Cooling off

Does the water cool faster than the sand? Use your terrarium and aquarium to find out.

Put your terrarium and aquarium in a pan of warm water. 1 Wait until the temperatures in the cups are the same. Then, take the cups out of the pan. Put both cups in a cool place. Watch the thermometers. Which cools first? 2 Were you surprised?

1

2

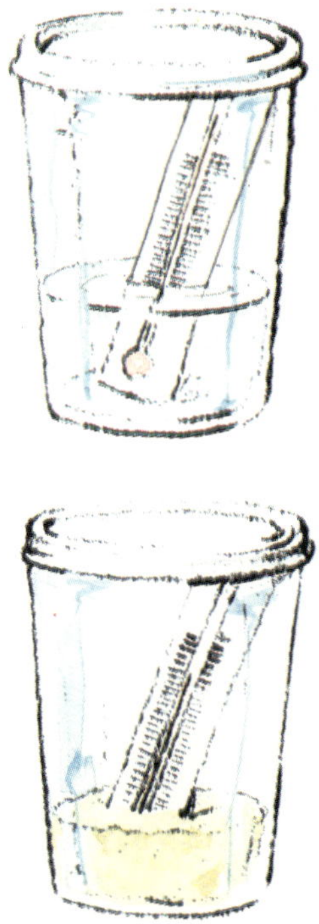

Deserts and oceans

This desert has lots of sand. 3
During the day, the temperature is very high. The sand heats quickly. Then, at night, the temperature drops. And the sand cools quickly.

3

Sand and air heat up faster than water. And they cool off faster than water.

The beach is a good place to go to cool off on a very hot day. 1 Why? The ocean water is cool. Even though the air and sand are hot. At night, the air and the sand cool quickly. But the ocean water feels warm!

1

Show what you know

Number your paper from 1 to 4. Read each sentence. Then write the word or words that best complete the sentence. Use the words in the Word Bag.

1. A terrarium is a small place for plants and ________ .
2. An aquarium is a small place for water plants and ________ .
3. Water ________ slower than sand and air.
4. Sand and air ________ faster than water.

Think it through

Should you put an aquarium in the sunniest part of your room? Why, or why not?

21

Push and pull

Up and down on a seesaw

What keeps the seesaws from going down on the left side? 1 Now, look at the picture of the boy. What would happen if the boy moved farther to the right? What would happen if he moved to the left?

1

The girl is **heavier** than the boy. She weighs 18 **kilograms** (about 40 pounds). He weighs 15 kilograms (about 35 pounds).

He is pulling down on the seesaw. 1 He cannot get his end down to the ground. The girl is pushing down 18 kilograms. He can pull down only 15 kilograms. What could she do to help him pull down his end of the seesaw?

1

The girl moved to the boy's side of the seesaw. 2 The boy pulled up on his end of the seesaw. The boy held his end off the ground. The girl moved closer to him. Then he had to pull up harder to keep his end off the ground.

2

How hard did the boy pull up when the girl was close? Did he pull up 18 kilograms?

Make a toy seesaw

Cut a piece of a straw. Make sure it is a little longer than the **width** of your ruler. Put a piece of clay the size of a nickel on your pencil. 1 Press the clay with a ruler to make the clay flat. 2 Take the pencil out of the clay. Put the piece of straw where the pencil was. 3

1

2

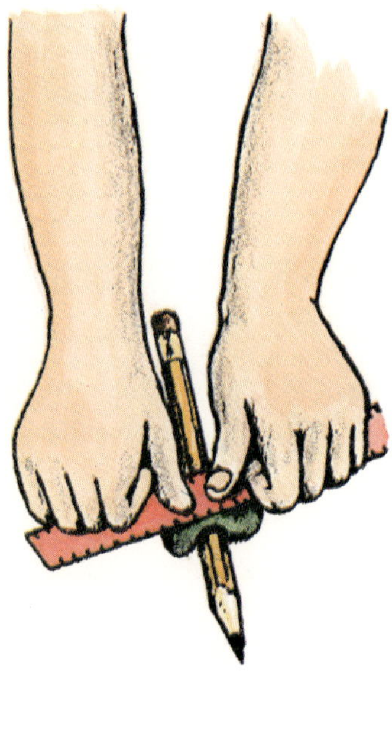

3

Put the clay and the straw underneath the ruler. 4 Make sure they are at the middle of the ruler.

Push the left end of the seesaw down. 5 Does it stay down? Push the right end down. 6 Does it stay down?

Do the ends of your seesaw stay down? Why, or why not?

4

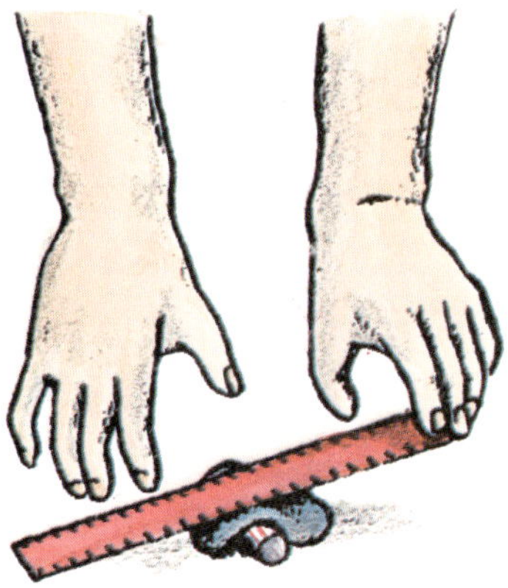

5

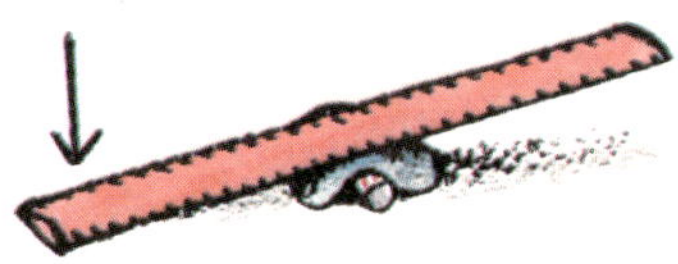

6

Activity

Blocks and drops

Put a piece of clay on one end of your toy seesaw. Stick a small cup on the clay. 1

Put two blocks on the other end of your seesaw. 2 Fill a **dropper** with water. Drop water into the cup until the cup side goes down. 3

Put another block on the two blocks. Drop more water in the cup. Count the drops needed to make the cup side go down again. Put another block on the seesaw. How many drops of water make the cup side go down now?

Make a chart. 4

1

2

3

4

Blocks	Drops of Water
1	30
2	30
3	28
4	30
5	30

Pushing up and down

The blocks push down on one end of the seesaw. 5 Water pushes down on the other end of the seesaw.

5

How many drops of water push up a block on your toy seesaw? What pushes down on the right end of your seesaw? What pushes down on the left end of your seesaw? What pushes up on your seesaw?

What pushes up on this seesaw? 6 What pushes down on this seesaw? Can the girl and boy push down harder than they push up?

6

Measuring push and pull

How hard is the girl pushing down? 1 Stand on a **scale**. What number does the pointer point to? That number is your weight.

Put a jump rope under the scale. Now stand on the scale. Pull the ends of the jump rope. 2 What number does the pointer point to now? Is the number greater than your weight?

1

2

Show what you know

Number your paper from 1 to 4. Write T for each sentence that is true. Write NT for each sentence that is not true.

1. The side of a seesaw with the most weight goes down.
2. It is not possible to pull up more than you weigh.
3. On a seesaw, you can push up harder when your feet touch the ground than you can push down.
4. A scale measures how hard you are pushing down.

Think it through

How might people use scales in their jobs?

Discover more

You can learn more about your world.

Ask your mom or dad to help.

Look at the animal.

Take good care of the animal.

1. Be sure it has food.
2. Be sure it is not too hot or too cold.
3. Be sure it has a clean home.

Grow plants.

You need: seeds, soil, water, sunshine, pot

Find leaves, shells, or rocks.

Write down what you see and find.

Vocabulary list

Picture credits

1: John R. MacGregor/Peter Arnold, Inc.; **2:** Alok Kavan/Photo Researchers; **4:** James Lester/Photo Researchers; **5:** Craig Aurness/Woodfin Camp; **6:** Randy Matusow/McGraw-Hill; **10L:** Robert Capece/McGraw-Hill, **10r:** Jerome Wexler/Photo Researchers; **11:** Robert Capece/McGraw-Hill; **12:** Manuel Rodriguez; **13:** Russ Kinne/Photo Researchers; **16:** Dick Rowan/Photo Researchers; **18:** Clifford Dolfinger/Photo Researchers; **19:** William H. Hodge/Peter Arnold, Inc.; **20:** Grant Heilman Photography; **21L:** Ray Hunold/Photo Researchers, **21c:** USDA, **21r:** Grant Heilman Photography; **22:** Randy Matusow/McGraw-Hill; **26:** Robert Capece/McGraw-Hill; **27L:** Kjell B. Sandved, **27r:** Grant Heilman Photography; **28tr:** P.W. Grace/Photo Researchers, **28L:** Manuel Rodriguez, **28r:** Grant Heilman Photography;

30: Yoav Levy/Phototake; **32:** Peter Vadnai/McGraw-Hill; **33-34:** Runk/Schoenberger/ Grant Heilman Photography; **36:** Robert Capece/McGraw-Hill; **38:** Shelly Grossman/ Woodfin Camp; **40:** William E. Townsend, Jr./Photo Researchers; **41t:** Kent & Donna Dannen/National Wildlife Federation Conservation, **41b:** Anthony Marcieca/Photo Researchers; **42:** Stephen J. Krasemann/Photo Researchers; **46:** WWF/Timm Rautert/ Bruce Coleman Limited; **48:** Kenneth W. Fink/Photo Researchers; **49:** China News Service/PPS/Photo Researchers; **50:** Jessie Cohen/Office of Graphics & Exhibits, National Zoological Park, Smithsonian Institution;

52: Jim Amos/Photo Researchers; **54:** Peter Vadnai/McGraw-Hill; **55:** Robert Capece/ McGraw-Hill; **56:** Arnout Hyde, Jr.; **57:** Robert Capece/McGraw-Hill; **58:** Jeffrey Fox/ Woodfin Camp; **59:** New York State Department of Health; **60:** Craig Aurness/ Woodfin Camp;

62: Randy Matusow/McGraw-Hill; **63-70:** Robert Capece/McGraw-Hill; **71:** Peter Vadnai/McGraw-Hill; **72:** Barbara Pfeffer/Peter Arnold, Inc.; **73:** Ann McQueen/Stock, Boston; **74:** Costa Manos/Magnum; **75t:** Shalmon Bernstein; **75b:** Rhoda Sidney/ Monkmeyer; **76:** Randy Matusow/McGraw-Hill; **78:** Deborah Feingold; **80:** Randy Matusow/Archive; **81L:** Katrina Thomas/Photo Researchers, **81r:** Randy Matusow/ McGraw-Hill; **83:** Cary Wolinsky/Stock, Boston; **84:** Nathan Ben/Woodfin Camp;

86-87: NASA/Woodfin Camp; **88:** Alan Pitcairn/Grant Heilman Photography; **90:** Robert Capece/McGraw-Hill; **91L:** (insert) Grant Heilman Photography, **92r:** Roger Maserang; **93:** Grant Heilman Photography; **98:** Manuel Rodriguez; **99t:** Tom McHugh/Photo Researchers, **99b:** Grant Heilman Photography;

102: Harvey Lloyd/Peter Arnold, Inc.; **104:** Robert Capece/McGraw-Hill; **107:** Farrell Grehan/Photo Researchers; **110:** Robert Capece/McGraw-Hill; **113:** Peter Arnold/Peter Arnold, Inc.; **114:** Clyde H. Smith/Peter Arnold, Inc.; **115:** Robert Capece/McGraw-Hill; **116:** Dan Sudia/Photo Researchers;

118: Alan Pitcairn/Grant Heilman Photography; **120:** Russ Kinne/Photo Researchers; **121t:** C. Knuckles/New Mexico State University; **121, 124:** NASA;

126-127: Harvey Lloyd/Peter Arnold, Inc.; **128:** Joan Liftin/Archives; **130-134:** Robert Capece/McGraw-Hill; **136:** Grant Heilman Photography; **138:** Manuel Rodriguez; **139:** Elizabeth Crews/Stock, Boston; **142:** Ken Lax/McGraw-Hill; **144:** Robert Capece/McGraw-Hill; **146:** Mike Mazzaschi/Stock, Boston; **147:** Elizabeth Crews/Stock, Boston; **150:** Ken Lax/McGraw-Hill; **151, 152:** Robert Capece/McGraw-Hill; **154:** Mike Mazzaschi/Stock, Boston; **156:** Ken Lax/McGraw-Hill; **159:** Robert Capece/ McGraw-Hill; **160:** Peter Vadnai/McGraw-Hill; **163:** Alan Pitcairn/Grant Heilman Photography; **164:** Farrell Grahan/Photo Researchers; **166, 168, 169:** Robert Capece/ McGraw-Hill; **171:** Alan Pitcairn/Grant Heilman Photography; **172:** Farrell Grahan/ Photo Researchers; **173:** Robert Capece/McGraw-Hill; **174:** Peter Vadnai/ McGraw-Hill.